001·94

THE UNEXPLAINED
MYSTERIES OF THE UNIVERSE

COLIN WILSON

DORLING KINDERSLEY

LONDON • NEW YORK • MOSCOW • SYDNEY

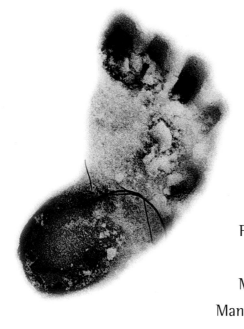

A DORLING KINDERSLEY BOOK

Project Editor *Amanda Rayner*

Art Editor *Alexandra Brown*

Managing Editor *Anna Kruger*

Managing Art Editor *Jacquie Gulliver*

Picture Research *Tom Worsley*

Production *Ruth Cobb*

DTP Designer *Nicola Studdart*

Designer *Robin Hunter*

First published in Great Britain in 1997
by Dorling Kindersley Limited,
9 Henrietta Street, Covent Garden, London WC2 8PS

Visit us on the World Wide Web at http:/www.dk.com

A CIP catalogue record for this book is available from the British Library.

ISBN 0 7513 5648 4

Colour reproduction by G.R.B., Italy
Printed and bound by L.E.G.O., Italy

ACKNOWLEDGEMENTS

The Publishers would like to thank: Karen Fielding for specially commissioned artwork; and the following for their kind permission to reproduce their photographs: *(c=centre; b=bottom; l=left; r=right; t=top, a=above)*

Academy of Applied Science, Robert H. Rines: 30cr; The Ancient Art & Architecture Collection Ltd: 5c, 11cla, 13ca, 23cb, br, 21cra; Barnaby's Picture Library: 27tl; Dr Alan Beaumont: 24bc; Beinecke Library: 21br; Bermuda Natural History Museum: Wolfgang E. Sterrer 8cl; Corbis UK Ltd: 24cla, 28crb; English Heritage Photographic Library: 14c; Robert Estall Photo Library: 17cra; Mary Evans Picture Library: 10cr, 12tr, 18cl, 23tr, 33tr, 35ca, 36tc, tr, cl; Fortean Picture Library: 7c, 17cb, 20cr, 29crb, 22c, 26cl, 30cl, 31cr, 33crb, c/ K. Aarsleff 12bc, 15tr, 19tr/A. Barker 32tc/ Booth Museum 19br/ J&C Bord 16tr/ R. Dahinden 6tl, 32clb, cb, crb, 33tc/ R. Le Serrec 31br/ F.C. Taylor 17c; Galaxy Picture Library: 29cr; The Robert Harding Picture Library: 11tr; Hebridean Press Service: 25cra; Michael Holford: 13br; Hutchison: 16clb; Images/ Charles Walker Collection: 14bc, 29cra, 27clb, 21cl; Frank Lane Picture Agency: 19tl; Magnum Photos: 14tr, 22cl, 23tc/ H. Gruyaert 21tc; Mirror Syndication International: 37tr; NASA: 35crb; Peter Newark's Historical Pictures: 25tl; Planet Earth Pictures: 31tr Peter Scones; Popperfoto: 22c; Science Photo Library: 20clb/ D.A. Hardy 28cl/ NCAR 26tr/ NASA 35tl, cr, cb/ P. Parviainen 4c, 27br, 29tl/ F.K. Smith 27cra/ Frank Zullo 34c; Scotland in Focus: 37ca; South American Pictures: 13tr, 15cla, 17tl; Spectrum Colour Library: 37tl; Frank Spooner Pictures: 30bc; Tony Stone Images: 36crb; The Werner Forman Archive: 18c, borders pp8-37, endpapers.

Jacket: Academy of Applied Science, Robert H. Rines: back tr; Mary Evans Picture Library: inside front tc; Fortean Picture Library: front cla, cra; Robert Harding Picture Library: front ca; Images/ Charles Walker Collection: front bc; Tony Stone Images: front cra, inside back c; Werner Forman Archive: front tc, back br.

Every effort has been made to trace the copyright holders. Dorling Kindersley apologises for any unintentional omissions and would be pleased, in such circumstances, to add an acknowledgement in future editions.

CONTENTS

INTRODUCTION

I have always been interested in unsolved mysteries. As a child I was fascinated by stories of the Loch Ness monster and the legendary curse of the pharaohs of Egypt. I envied archaeologists who could visit the sites of ancient cities that lay buried under the earth – for they could uncover them to learn the secrets of people who lived thousands of years ago.

The Piri Re'is map may show the South Pole before it was covered with ice

When I was ten, my grandfather gave me a book about the world's greatest wonders. In that book I found pictures of Stonehenge, the mysterious circle of stones in Britain. Even today no one knows for certain why it was built. And I was

Dozens of planes vanished without a trace...

Avenger torpedo bomber – lost in the Bermuda triangle

Bermuda Blob – the remains of a giant octopus found in 1997

amazed to read about the Yeti, a strange, hairy creature that is believed to live in the mountains between India and Tibet. Other giant creatures, like the Bermuda Blob, lurking in the deepest oceans, are rarely seen by humans but we have photographs to prove that they exist.

In modern times, experts continue to be baffled by objects such as the Piri Re'is map, left to us by ancient people. We may never know who made these treasures or discover their secret purpose.

Many mysterious events are still reported from all around the world – statues of the Virgin Mary (left) have cried tears of blood.

And the peculiar disappearance of dozens of aeroplanes in the "Bermuda triangle" may be evidence of great leaps in space and time. In fact, weird happenings like these have been reported for hundreds of years and are perhaps the most extraordinary puzzles of all.

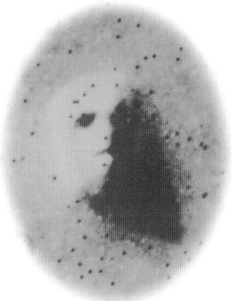

Through space travel we are slowly finding out more about nearby planets. On the surface of Mars the outline of a face was recently photographed – no one knows how the shape was formed. The latest amazing object to become famous is a meteorite from Mars. Some scientists believe that this lump of rock shows there was once life on Mars. When humans finally land there, we may find the answer to this question. In the meantime, thousands of other mysteries are still waiting to be solved...

Photograph showing the shape of a face on Mars

Colin Wilson

Stonehenge is one of the most mysterious places on Earth

ANCIENT MYSTERIES

There are many strange monuments on the Earth that seem to have magical powers. Everyone who visits the pyramids and the Sphinx in Egypt senses their mysterious qualities. Intriguing discoveries have convinced some archaeologists that our distant ancestors had far more technical knowledge than we give them credit for. However, to this day no one knows how ancient people built the Great Pyramid.

Were the Pyramids Built by Aliens?

A minority of people believe that the pyramids were built by visitors from another planet. Perhaps god-like beings passed on their skills to help early humans build their great monuments.

Pyramid-like objects and a Sphinx-like face have been spotted on the surface of Mars (see page 9). Some people now think that the same extraterrestrials made the Egyptian monuments. Interestingly, the features of the Sphinx and the Martian face appear to have some similarities.

The face of the Sphinx was modelled on a real pharaoh, possibly Chephren who ruled 2556–2530 BC

Ancient people said that there were secret chambers under the Sphinx – these are now rumoured to have been found

THE SECRET OF THE SPHINX

The age of the Sphinx is one of the world's great mysteries. Its features have been worn away and some archaeologists believe it was seriously damaged by water many centuries ago. Was the Sphinx once buried at the bottom of the sea? The damage looks too severe to be the result of wind-blown sand. Some experts now think that it was built earlier than 2500 BC.

Lost People

Human memory is short, and many great civilizations have vanished and been forgotten. With every new discovery archaeologists can find out a little more about past people and places. The lost Indus city of Mohenjo-Daro was destroyed by some unknown disaster about 1900 BC. It was rediscovered by accident in 1922. Amazing statues were unearthed that tell us something about how the people must have lived. Objects left behind by the Etruscan people from Italy, who disappeared more than 2,000 years ago, have provided some clues about their everyday lives. However, we still do not know the full story…

Soapstone statue of a priest from Mohenjo-Daro

Etruscan painting found in a tomb in Italy

The truth is hidden in the stars....

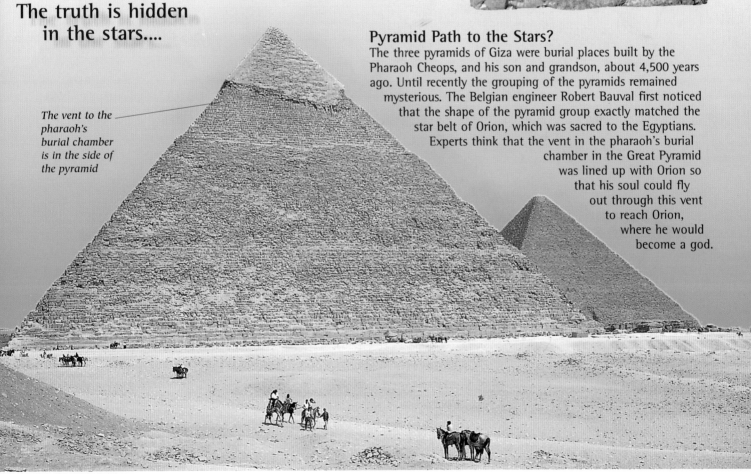

The vent to the pharaoh's burial chamber is in the side of the pyramid

Pyramid Path to the Stars?

The three pyramids of Giza were burial places built by the Pharaoh Cheops, and his son and grandson, about 4,500 years ago. Until recently the grouping of the pyramids remained mysterious. The Belgian engineer Robert Bauval first noticed that the shape of the pyramid group exactly matched the star belt of Orion, which was sacred to the Egyptians. Experts think that the vent in the pharaoh's burial chamber in the Great Pyramid was lined up with Orion so that his soul could fly out through this vent to reach Orion, where he would become a god.

LOST WORLDS

The Earth is full of lost places. The mythical island of Atlantis is said to have disappeared beneath the waves about 9500 BC, and Lemuria (in the Pacific Ocean) vanished at about the same time. The Inca city of Machu Picchu remained hidden for almost four centuries before it was rediscovered in 1911. Even the whereabouts of Camelot, the mysterious place where King Arthur held his court, is still unknown.

MACHU PICCHU – SECRET CITY

In the Andean mountains of Peru, in 1532, the Spanish conqueror Pizarro brutally strangled the Inca king Atahualpa. The king cursed his murderers, and they all died. The king's brother, Manco Capac, fought the Spaniards, but when they won, he withdrew to a secret hiding place, Machu Picchu, perched on a mountain top. His enemies never managed to find him. Eventually, the city was deserted and forgotten. It was found by the American explorer Hiram Bingham, who was led there by a native South American, in 1911.

A painting showing how the palaces of Atlantis may have looked

Atlantis – the Land Beneath the Waves

The philosopher Plato first told the story of Atlantis in about 350 BC. He said that it was a great island in the Atlantic Ocean that had vanished beneath waves 9,000 years earlier. Since then there have been hundreds of books about it. The latest theory is that Atlantis was the continent now called Antarctica, and that the remains of great cities now lie deep under the ice.

Machu Picchu was once home to 10,000 people

El Dorado – The Man of Gold

When Spanish explorers arrived in Venezuela in 1529, they noticed the gold ornaments worn by the native South Americans. This gold, they heard, came from "El Dorado", in the mountains. They assumed this meant "Land of Gold", and many explorers died seeking this fantastic place. But when the source of the gold was finally found, in 1536, it was only a small village near Lake Guatavita, Colombia. The local people had grown wealthy by trading. When a new king was crowned, he was covered in gum, then coated with gold dust. The king floated on the lake, gold ornaments were thrown in after him as an offering, and he swam until the gold washed off. It seemed that El Dorado did not mean Land of Gold – but "Man of Gold".

Lake Guatavita, Colombia

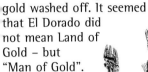

Golden model of a raft carrying the new king

Camelot and the Round Table

King Arthur's knights sat at the Round Table in Camelot in about AD 500. Arthur's legendary capital was a great castle on a hill. Experts think that the site of the capital is Cadbury Castle, an ancient hill-fort south of Glastonbury in western Britain. Local village names like Queen Camel and West Camel suggest that they may be right. It is said that King Arthur sleeps in a cave in the hill, and that he will one day wake up and return.

King Arthur sleeps. One day he will return...

Replica Round Table, made about 1340, now in Winchester, UK

STONE POWER

Giant stone circles and megaliths (big stones) exist all over the world, although no one knows quite how or why they were erected. But we know that the Earth is a huge magnet, with magnetic forces influenced by the Sun, Moon, and the planets. And the Earth is alive, just like a human body. One theory suggests that the stones were placed in the Earth to release concentrations of "Earth force" and spread fertility across the land.

Carnac stones in Brittany, France

Carnac

One of the most baffling sites in the world, Carnac, on the coast of Brittany in France, has 11 rows of standing stones, called menhirs. These stones stretch for 6.5 km (4 miles) and there are more than 3,000 of them. The oldest date from 7,000 years ago. We know they served some religious or magical purpose, but today no one has the slightest idea what this was.

STONEHENGE – THE FIRST COMPUTER?

A computer expert has shown that Stonehenge in Britain was a great stone calendar, whose megaliths mark the rise of the Sun and Moon over the centuries. In other words, Stonehenge was nothing less than a highly sophisticated computer, and the people who designed it were, as one expert put it, "Stone-Age Einsteins".

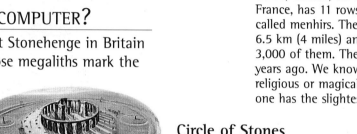

Painting of Stonehenge, UK

Circle of Stones

Stonehenge was once called the Giants' Dance, and there is a legend that it was built by giants. In fact, we now know that it was started about 5,000 years ago as a temple, and that its priests performed ceremonies to the Earth Goddess to persuade her to grant a good harvest.

The lintel stones were laid across the top

The sarsen stones form an upright ring

Island of Giants

For more than two centuries after their discovery by the Dutch in 1722, the origin of the Easter Island people and their giant statues was a mystery. The islanders said that the statues had walked there. However, the islanders' ancestors came from far over the sea in boats or rafts and found a fertile island covered with forests. For centuries they lived well and carved great statues of their gods from volcanic rock. But gradually they destroyed their forests and newcomers to the island brought deadly smallpox. Now all that remains are the stone statues keeping watch over the empty island.

Restored stone statues, called *moai*, with red wig-stones

Splendid isolation – a view from Easter Island

The stones were arranged to make a huge astronomical calendar

Who Built the Stone Circles?

Even today no one knows who created the stone circles or how they did it. Was it:

• Ancient people who used machines to move the heavy stones?

• Spirits or mysterious giants who lifted the stones by magic?

• Priests who organized their construction for centres of worship?

• Aliens who built the circles as landing areas for their spacecraft?

EARTH MYSTERIES

The Earth is full of strange forces that even modern science does not understand. Why did the North and South Poles change places 30,000 years ago? When will it happen again? What causes the Earth's magnetism? What was there about certain hills or mountains that made them sacred to our ancestors? People who are alive today have surely forgotten many things about the Earth that our ancestors understood.

Ley line in Wiltshire, UK

Ley Lines

Ancient humans believed that long lines of force that may be linked to magnetism ran across the countryside, and that in places this force became concentrated into a kind of whirlpool. Such places were regarded as sacred, and temples (and later churches) were often built on them. Stonehenge stands on such a site of mysterious energy. We now call these lines of force "ley lines".

ULURU

Uluru, or Ayers Rock, is a huge sandstone mound in central Australia that has been sacred to the Aboriginal people for hundreds of years. They believe that it was made by spirits in the "Dreamtime", the time when the world was created. The Aboriginals also have an equivalent of "ley lines", which they call "dreaming tracks". These tracks join sacred places, and because each one has hundreds of stories and songs linked to it, they are also known as "song lines".

A Rock is Born

Many aboriginal paintings (above) tell important Earth stories. One legend states that Uluru was originally a lake. However, after a terrible battle, the Earth rose up in revolt at the bloodshed, to form the great blood-coloured rock.

A Bird's Eye View

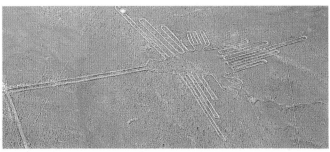

Lines in the shape of a hummingbird, Nazca Plain, Peru

Desert Drawings

The Nazca lines are huge "drawings" – often of birds or animals – made on the desert surface in Peru by removing stone after stone to form a gigantic picture. Because rising hot air in this hot, dry landscape prevents wind from blowing at ground level, the lines have been preserved for a long time, possibly 2,000 years. No one understands quite how or why they were made.

The World's Biggest Snake

The serpent mound of Ohio is one of hundreds of earth mounds found in the Midwest of the United States. It is 366 m (1,200 ft) long and was probably built by Native Americans – possibly the Adena people – in about 600 BC. Such mounds may be religious sites, and were often used for the burial of the dead. Temples were built on top and many were surrounded by wooden walls.

Uluru... the great blood-red rock

Aliens Squash Corn!

Crop circles seen in Westbury, Wiltshire, UK, in 1988

Crop circles were spotted in Wiltshire in western Britain in August 1980, when three perfect circles of flattened oats appeared in a farmer's field. The circles have since been seen all over the world and have become much more complicated.

How are they made?

Some people think that crop circles are made by UFOs. Various hoaxers have confessed to making some of the circles, but others seem to have been formed naturally. Witnesses claim they have seen a kind of whirlwind, making a high-pitched whining noise, actually producing the circles.

"If this is E.T. ringing home, I hope he uses someone else's corn field next time."
A farmer, UK, 1990

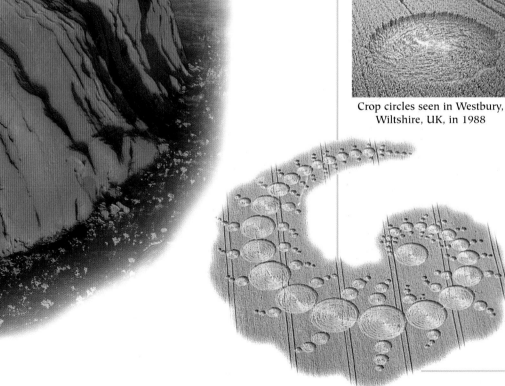

Crop circles seen near Stonehenge, Wiltshire, UK, in 1996

CURSES AND CURIOSITIES

Humans are sometimes called "tool-making animals", and ancient people were skilled in many crafts. A hundred thousand years ago, the ape-like Neanderthals made perfectly round discs. Their purpose is unknown, although they are thought to represent the Sun and Moon. Only one thing is clear – these and other mysterious objects, like granite balls, glowing stones, and animals carved from wood, served some kind of magical purpose.

Some spook is keeping watch over the dead pharaoh...

The tomb of Tutankhamen is opened by Howard Carter and Lord Carnarvon, 1922

CURSE OF THE PHARAOH

The tomb of the Egyptian pharaoh Tutankhamen was opened by Howard Carter and Lord Carnarvon in November 1922. It was found to be full of treasure, including a magnificent golden burial mask. An inscription in the tomb warned that "Death will slay with his wings whoever disturbs the peace of the pharaoh". Lord Carnarvon died soon afterwards – some people said he was killed by the pharaoh's curse.

The moving stones of Death Valley, California, USA

Death Valley Stones

Death Valley, in the USA, has many dried lake beds, called "playas", of smooth clay. Enormous stones, some weighing more than 50 kg (110 lb), frequently leave long tracks behind them, showing that they have moved. A study by geologist Robert P. Sharp from 1969 to 1976 found that when the clay becomes sticky and slimy in winter, strong winds can blow the stones long distances, like sleds slipping over ice.

A granite ball measuring about 2.5 m (8 ft) across, found in Costa Rica

Mystery Balls

When the United Fruit Company cleared the jungle in the Diquis Delta in Costa Rica, Central America, in the 1930s, they found stone spheres ranging from a few centimetres to 2.5 m (8 ft) across. There are probably thousands of these balls in the jungle, and the granite from which they are made is not found naturally in that area. No one knows who carved them or why, although it seems likely that they represent the stars and planets.

The booya stones glow with a secret inner power

X-ray Stones

In the Murray Islands, north of Australia, the priests guard three sacred stones, known as the "booya stones", that glow with a fierce blue light. The Australian writer Ion Idriess stated that when the light is focused into a beam and directed at a human being, it causes an X-ray effect that always results in death. When Europeans came to the Murray Islands, the priests hid the stones, and no one knows where they are now kept.

Toad in the Hole

Death-defying Toads

In the Booth Museum, Brighton, UK, there is a mummified toad that was found sealed inside an egg-shaped flint. And in 1865, a living toad was found by workmen in Hartlepool, UK, inside a block of magnesium limestone 7.5 m (25 ft) under the ground. How is this possible? Toads have an amazing ability to survive inside the dried mud of ponds – some have been known to live for as long as 12 years in these conditions.

Toad encased in stone, found in 1901, now kept in the Booth Museum, Brighton, UK

MESSAGES FROM BEYOND

In the last 20 years, archaeologists have pushed back the dates for the beginning of human society. Discoveries like the Piri Re'is map and the Antikythera computer suggest that our distant ancestors knew more about the world and about technology than even the boldest historian has claimed.

SKULL OF DOOM

According to the adventurer Mike Mitchell-Hedges, his daughter Anna found a crystal skull in a Mayan temple in Central America in 1927. He claimed it was 3,600 years old and had the power to strike people dead, although some people believe that he bought the skull in London, UK, in 1944. The life-size skull does appear to have "spooky" qualities. When a Californian scientist, Frank Dorland, studied the skull in his house, objects moved by themselves and strange sounds were heard all through the night.

Crystal skull said to have been found in Central America in 1927

Ancient computer found near Greece in 1900

Crusty Computer

In 1900, a diver near the Greek island of Antikythera found a badly worn scientific instrument with cogged wheels. It was discovered in a ship that had sunk in about 65 BC. In 1959, Professor Derek de Solla Price announced that it was a computer for finding your way at sea. It seems incredible that an unknown Greek genius invented such a complex machine more than 2,000 years ago.

"...We spotted something shining through the stones."

Anna Mitchell-Hedges

Weird Words

Rongo-rongo, the Easter Island language

We would know much more about Easter Island and where its people came from if we could read its language, rongo-rongo, which is carved on wooden tablets. Sadly, the secret is lost. In 1862, a Peruvian slave ship kidnapped most of the islanders, and took them to Peru to work in the mines. Most of them died, and the few who were finally sent home brought the deadly disease smallpox, which killed most of the others. There is no one left alive today who can read rongo-rongo.

This clay tablet was found in the ruins of the ancient palace in Phaistos, Crete. It is roughly circular in shape, 16.5 cm (6.5 in) across, red, and covered on both sides with mysterious signs. No one has been able to make sense of these signs, although the British archaeologist Sir Arthur Evans guessed that it is a poem written about an important battle.

The Phaistos disc, found in Crete

Magical Map

Piri Re'is was a Turkish pirate, beheaded in 1554, and a map made by him was discovered in Istanbul in 1929. It not only shows the coast of South America, but the South Pole as it was before it was covered with ice, perhaps 8,000 years ago. It has been suggested that the map was made by "space visitors" who came to Earth long before human life. It certainly seems to prove that there were great sea-travellers now long forgotten by history.

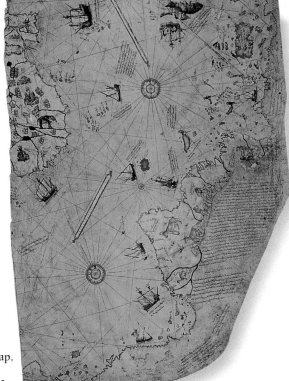

Piri Re'is map, drawn in about 1513

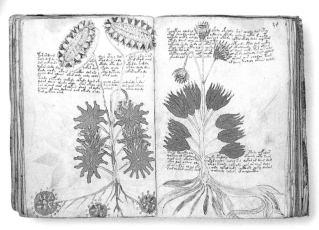

Voynich manuscript, kept at Yale University, USA

Baffling Book

Wilfred Voynich bought this strange book in 1912. It had 204 pages with pictures of flowers and diagrams of the stars, and was written in code. A letter found with it, in a monastery in Frascati, Italy, was dated 1666. All attempts to break the code have failed.

HOLY MYSTERIES

Unlike ancient maps or computers, some mysterious objects do not seem to have a practical use. The strange qualities of the Turin shroud have puzzled experts for hundreds of years. And although some weeping statues or paintings may be fakes, there are far too many for that explanation to apply every time. Some mysteries may very well be miracles.

Christ Weeps in Bethlehem

On 28th November 1996, in the Church of the Nativity, Bethlehem, an image of Christ appeared to be crying real tears. Many Christians believe he was weeping for the sins of the modern world.

"I see his eyes open and close… There is no doubt in my mind that it is a miracle."
Father Anastasios, priest, Bethlehem

Weeping Christ in Bethlehem, West Bank, 1996

Worshipper feeds a statue of Ganesha, South London, UK, 1995

Milky Miracle

On 21st September 1995, a statue of the elephant-headed Hindu god Ganesha began drinking milk offered by worshippers in New Delhi, India. Soon this "miracle" was being reported all over the world. The following day, it stopped as suddenly as it had started.

"When we put a spoon of milk up to the lips… the milk disappeared in seconds."
Roshan Lal Bhanbari
Vishwa Temple, Southall, UK

TEARS OF BLOOD

In 1953 at Syracuse in Sicily, a plaster model of the Virgin Mary began to weep. Scientific tests showed that the tears were real. Other statues have wept tears of human blood. This effect could be faked if the statue had a small hollow inside its head and was filled with blood or tears, but so far, no trickery has been detected.

A statue of the Virgin
Mary cried tears of blood
in September 1982

Ship's fossilized timbers, Mount Ararat, Turkey

Is This Noah's Ark?
There is much historical evidence for great floods like the one described in the Bible. According to the Bible, Noah's ark landed at Mount Ararat, in Turkey. In 1984, an American-Turkish team found the outline of a ship on the side of Mount Ararat.

Supernatural Powers
The Ark of the Covenant (represented above) was a sacred chest that the ancient Hebrews carried into battle. It vanished centuries before the birth of Christ. Many Christians believe that this "lost ark" is now kept in a chapel in Axum, Ethiopia. It is said to have the power to cause illness and even death.

Fantastic Fake

The Turin shroud, a sheet of cloth showing the faint brown image of a man, has been known since 1353. It is now in Turin Cathedral, Italy. In 1898 a photographer was amazed to discover that the shroud is a "negative", and that through reversing it, he had turned it into a real photograph. Was this the outline of the body of Christ, miraculously fixed onto the cloth? Unfortunately, it wasn't. Recent tests show that the image was painted on the cloth.

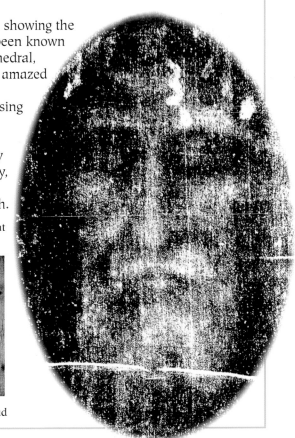

This image was once thought
to be the face of Christ

Full-length photograph of the Turin shroud

STRANGE DISAPPEARANCES

Disappearing ships, disappearing aeroplanes, disappearing people – these are perhaps the most alarming mysteries of all. Of course, many of them have perfectly simple explanations – ships sink in storms and aeroplanes crash through mechanical failure. But there are still unexplained disappearances that continue to amaze and baffle us.

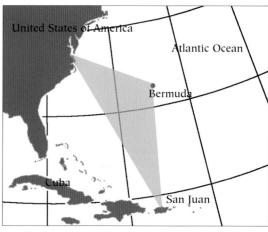

Map showing the Bermuda triangle area

Ships That Slipped Away

The sea around Bermuda is the scene of many disappearances. Between 1950 and 1984, nine ships simply vanished. The *Marine Sulphur Queen* (above) disappeared without any clues on 2nd February 1963, while travelling from Texas to Virginia. Thirty nine people were lost.

BERMUDA TRIANGLE

On 5th December 1945, five Avenger torpedo bombers took off from Fort Lauderdale, Florida, USA, on a routine flying mission. A few hours later, the Flight Leader radioed for help. A Martin Mariner flying boat went to look for them. None of the planes ever returned. Fifteen more planes disappeared in a triangular area around Bermuda over the next 30 years. It is suspected that some vortex – or "whirlpool" – of magnetism interferes with their navigation equipment so they become lost. None of the crews has ever been found.

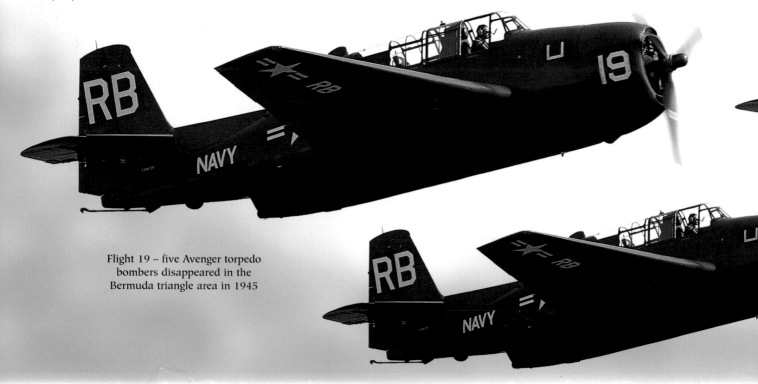

Flight 19 – five Avenger torpedo bombers disappeared in the Bermuda triangle area in 1945

The *Mary Celeste*

On 5th December 1872, the *Dei Gratia* spotted another ship, the *Mary Celeste,* drifting in the North Atlantic. When the crew climbed aboard, they found it deserted. The lifeboat was missing, and the ship had been abandoned in a hurry. Three barrels of its cargo of crude alcohol had burst open. It seems likely that the captain had ordered everyone into the lifeboat, afraid that the alcohol was about to explode. No bodies were ever found.

Artist's impression of the crew of the *Dei Gratia* rowing towards the *Mary Celeste*

Eilean More lighthouse, Scotland, UK

Lighthouse of Lost Souls

When the light of the Eilean More lighthouse, Scotland, UK, was seen to be unlit on 26th December 1900, a ship went to investigate. The three lighthouse keepers were missing and the last entry in their log book was for 15th December, when the weather was good. The mystery may have been solved in 1947, when a visiting journalist saw the sea heave up 20 m (70 ft) for no obvious reason. Perhaps a freak wave had dragged the keepers into the sea in 1900.

Do the planes enter a gateway to another world?

Bermuda Disappearances

1880 – British ship, the *Atlanta,* vanished with 290 people on board

30th October 1954 – Navy Lockheed Super Constellation plane vanished with 42 people on board

11th January 1967 – Cargo plane lost with 4 people on board

3rd June 1984 – 27-m (88-ft) ship, the *Marques,* lost with 18 people on board

and many, many more…

WEIRD WEATHER

Waterspouts that can sink a ship and giant hailstones that can smash through roofs – these are examples of the terrifying power of the weather. Our ancestors believed that storms were caused by angry gods and even today we cannot begin to control these immense natural forces.

Giant hailstone that fell in Kansas, USA, in 1970

Huge Hailstones

Water in rain clouds can remain liquid even when its temperature is well below freezing. However, when this water touches a speck of dust, it suddenly turns to ice. Tossed around inside the cloud, the pieces of ice can keep growing until they are larger than tennis balls. Finally, they get so heavy that they must fall to Earth – sometimes smashing windows and injuring people. This hailstone (above) measured 15 cm (6 in) across and weighed 765 g (1.7 lb).

IT'S RAINING FISH!

Showers of live fish, frogs, tadpoles, lizards, worms, and other creatures have been reported for hundreds of years. High winds, or perhaps whirlwinds, may sweep things into the air, hold them up, then drop them back to Earth. Rain, hail, snow, or fish will fall to Earth sooner or later because what goes up must come down. In 1989, it rained sardines on Ipswich, Australia, during a violent storm. No one knows where mystery fish will land next...

This picture of falling fish was based on an ancient engraving

Photograph of the River Thames, London, UK, 1890s

Mini Ice Ages

During the 20th century, the climate in northern Europe has been fairly mild, but not so long ago there were "little ice ages". Between 1550 and 1800, the winters in the UK were often so cold that "ice fairs" were held as entertainment on the River Thames in London. Even in the 1890s, it was cold enough to partly freeze the river, as you can see in this photograph of the River Thames (left).

Waterspout and lightning seen over Florida, USA, 1992

Water-loving Dragons?

Hot air rises – often so fast that it whirls around, forming a whirlwind. If water vapour is sucked up over the sea, it turns into a waterspout. These are sometimes seen at the same time as dramatic streaks of lightning. Sailors are terrified of them and ancient Chinese people believed that dragons caused waterspouts at sea and whirlwinds on land.

Amazing Balls of Light

"Ball lightning" may occur during violent thunderstorms. It is a glowing ball of red, yellow, or white light that floats like a balloon and can even pass through walls. In Paris, France, in July 1852, a ball the size of a human head came down a chimney, drifted across a room, then went back up the chimney and exploded, causing great damage. The balls are believed to be made of plasma – a kind of highly-charged gas. Remarkably, ball lightning does not give off any heat.

Coloured engraving of ball lightning, 1866

Heavenly UFOs?

Rays of sunlight appear bent when they pass through thin clouds. These rays sometimes seem to form coloured rings. This type of ring around the Sun (right) is a called a "corona", which means crown. Similar shapes can often be seen around the Moon. Such haloes around the Sun and the Moon have fooled some people into thinking they have seen a UFO.

Corona photographed over Finland, 1992

STRANGE PHENOMENA

Mysteries of the natural world remind us of how little we really know about our planet. Weird phenomena such as earthlights and the comet Hale-Bopp give rise to much speculation. Scientists all over the world are studying these wonders, hoping to unlock their secrets. Now we are just beginning to grasp the enormity of the natural disaster that killed off all the mammoths and sabre-toothed tigers.

ICE AGE DEATH TRAP

La Brea, in Los Angeles, USA, has a lake of asphalt, a black, sticky substance like tar that rises to the surface from oil-bearing layers of rock. The bones of thousands of ancient animals dragged down to their death have been thrown up by this sticky pit, including mastodons, mammoths, sabre-toothed tigers, giant sloths, and giant camels. Many of these animals lived 40,000 years ago during the last ice age, when a great disaster seems to have rocked the American continent from north to south, killing off dozens of species of animals and plants.

Model of a mighty mastodon trapped by tar at La Brea, USA

Painting of the end of the dinosaurs

Death of the Dinosaurs

Dinosaurs vanished from the Earth about 65 million years ago. Now scientists believe that a giant asteroid, measuring 20 km (12 miles) across, fell to Earth near the Yucatan Peninsula in Mexico. The asteroid hit the Earth at thousands of kilometres per hour, hurling billions of tonnes of dust and water vapour into the atmosphere. This blocked out the sunlight and caused a 5,000-year winter that killed off most plants and animals.

Bright Sparks

The northern lights, or aurora borealis, light up the skies in the far north of the world in a spectacular display. The lights form patterns such as "curtains", "rays", and "crowns". They were once believed to be a warning that a disaster was about to happen, but they are now known to be charged particles from the Sun striking against our atmosphere. These particles are attracted towards the North and South Poles. At the South Pole the effect is known as the southern lights.

Northern lights photographed over Finland

Christ's face in the clouds, Korea, 1950s

Face in the Clouds

During thunderstorms, clouds can take on strange shapes such as arches, doughnuts, balloons, cigars, globes, and even flying saucers. During the Korean War (1950–53), pilots were astounded to see a vision of Christ in the clouds above Korea.

Balls of Fire and Ice

Aliens Tail Hale-Bopp!

The comet Hale-Bopp (above) appeared in the sky in March 1997. It was bright enough to be seen without a telescope. Like all comets, it is a giant ball of ice. Many people claimed they saw a shining object near Hale-Bopp surrounded by a ring like the planet Saturn. They believed it was a spaceship heading towards the Earth and aliens on board were using the comet to hide their investigation of human activities.

Spooky Earthlights

In areas where earthquakes are common, balls of fire called earthlights are often seen. Most scientists believe that they are produced when the rocks of the Earth are under pressure. These lights sometimes streak across the sky and may explain some UFO sightings.

"They appeared to be large and round, and some were quite bright."
Jim Howard, Texas, USA

Earthlight photographed in New Jersey, USA

MONSTERS OF THE DEEP

Most people assume that dinosaurs vanished from the Earth millions of years ago. However, it now seems possible that a few have survived. The famous Loch Ness monster may be a living dinosaur-like reptile called a plesiosaur. And many people have found evidence of giant squids that lurk in the deep seas. Dozens of unknown monsters may live in remote lakes and the unexplored depths of the oceans.

Underwater photograph taken in Loch Ness, Scotland, UK, in 1975, that may show the monster

LOCH NESS MONSTER

People began to report sightings of the Loch Ness monster, or "Nessie", in April 1933, when a new road was built on the north shore of the loch in Scotland, UK. A local couple, Mr and Mrs Mckay, saw a huge creature with two black humps swimming across the loch. Two more people saw a strange animal crossing the road with a sheep in its mouth. This underwater photograph possibly showing the monster was taken in 1975. Although there is now a Loch Ness Investigation Bureau, many scientists insist that the Loch Ness monster is a fantasy.

The surgeon's photograph, taken in 1934

The Surgeon's Photograph
The most famous image of the Loch Ness monster was taken in 1934. The "surgeon's photograph" was declared to be a hoax 50 years later. It was a photograph of a plastic "neck" attached to a toy submarine. The surgeon, Robert Wilson, claimed he took the picture as "a bit of harmless fun".

Champ of the Lake
One of Nessie's cousins may live in Lake Champlain on the US/Canadian border. The shy beast, nicknamed Champ, was photographed by Sandra Mansi in 1977. Despite more than 250 sightings, no one knows what type of creature it is.

Photograph of a creature in Lake Champlain, USA/Canada, 1977

Living Fossil

The coelacanth is a powerful fish about 1.5 m (5 ft) long. Scientists thought that it had become extinct 60 million years ago, but in 1938 a fisherman caught a live coelacanth in the Indian Ocean, off the coast of Africa. Scientists call the coelacanth a "living fossil".

Is Nessie a living dinosaur?

Coelacanth photographed near the Comoros Islands, Indian Ocean

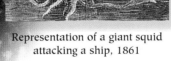

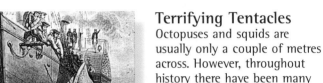

Representation of a giant squid attacking a ship, 1861

Terrifying Tentacles

Octopuses and squids are usually only a couple of metres across. However, throughout history there have been many reports of creatures with tentacles long enough to pull a ship underwater. In 1896, a dead octopus was found on a beach in Florida, USA, with tentacles over 9 m (30 ft) long. As recently as 1984, a ship's captain from Bermuda was sure that a giant octopus had towed his trawler along by pulling on one of its fishing lines.

Monster Tadpole

This strange sea creature was spotted by photographer Robert le Serrec on 12th December 1964. He was out in his boat near Hook Island in Australia's Great Barrier Reef. The monster looked like a giant tadpole, about 20 m (65 ft) long, and seemed to have a wound in its side. When it raised its head, Robert le Serrec fled in terror.

Monster photographed near Hook Island, Australia, 1964

SCARY MONSTERS

Do creatures like the Yeti, Bigfoot, and the Beast of Exmoor really exist, or are they just fairy stories? Bernard Heuvelmans, a famous detective of the animal kingdom, believes that the world is full of creatures still unknown to science. Some may live in deserts or great tropical forests and others in the frozen wildernesses of the Earth. The "Minnesota Iceman", a creature found in a block of ice, seems to prove his point.

Wild cat found at Kellas, Scotland, UK, in 1983

Killer Cats

Early in the 1960s, sightings of huge wild cats were reported all over the UK and in the 1980s, a creature that killed sheep by ripping out their throats became known as "the Beast of Exmoor". Witnesses said it looked like a puma. There have been hundreds of similar sightings of huge, black panther-like creatures in the USA.

Bigfoot

The first reports of Bigfoot, a giant ape-like beast, date back to 1811 when a trader reported seeing 35-cm (14-in) long footprints. A logger called Albert Ostman claimed that he had been kidnapped by a Bigfoot in 1924, and was held prisoner by a family of four creatures until he managed to escape. The pictures below are actual stills from a film taken in 1967.

In 1967, Roger Patterson filmed what he claimed was a Bigfoot in Bluff Creek, California, USA. He and his friend Bob Gimlin thought that the creature was a female.

The creature turned around and looked at them. But as the men started to follow it, the Bigfoot began to run, and they soon lost its tracks in the pine needles.

"They looked just like Bigfoot is supposed to, hairy, huge hands and very powerfully built."

Glenn Thomas, Estacada, Oregon, USA, who saw three huge creatures

MINNESOTA ICEMAN

The Iceman was a hairy, human-like creature kept frozen inside a block of ice. In the 1960s, it was put on show in the Midwest of the USA as part of a travelling circus. The showman, Frank Hansen, stored the creature in a refrigerator. Many people who saw the Iceman believed that it was a well-preserved early human, but others thought it was a fake.

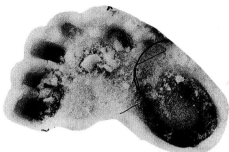

Freaky Footprint

A strange footprint (above) was found in the snow at Bossburg, Washington, USA in 1969. It was 42 cm (16.5 in) long. Experts believe that the print was made by a large ape-like creature, possibly a Bigfoot.

Artist's impression of a Yeti

Yeti

The first European travellers in Tibet were frightened by reports of a huge creature called a Yeti or Abominable Snow Man (because it had a horrible smell). In 1951, the Everest explorer Eric Shipton saw possible evidence. He found giant footprints in the snow that were 45 cm (18 in) long.

Mysterious frozen body that went on show in the 1960s

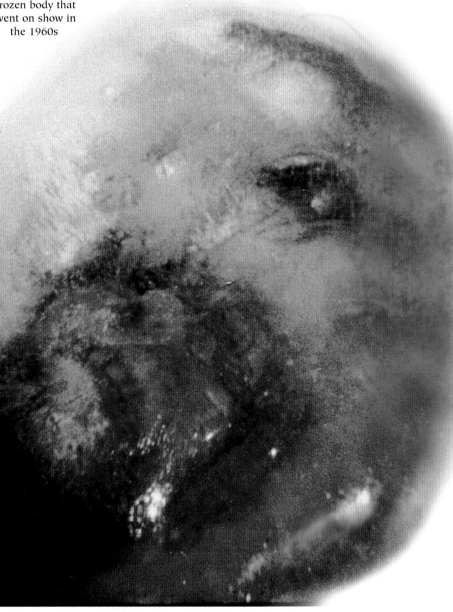

Loys's Ape

In 1920, a Swiss geologist, François de Loys, was exploring the Colombian/Venezuelan border when two tall, monkey-like creatures ran towards him screaming with rage. His men opened fire and killed the female, then photographed the body (above). The animal seems to be an unknown type of ape with a human-like face.

THE FINAL FRONTIER

Looking at the vast number of stars in the night sky, it seems almost certain that there is life somewhere out there. No living things were found on the Moon, although there may once have been some sort of life on Mars. Now scientists all over the world are working together on a major project called CETI (Communication with Extra Terrestrial Intelligence). They are trying to attract the attention of "other worlds" by beaming radio signals out into space.

There are at least two hundred billion stars in the Milky Way

IS ANYONE OUT THERE?

There is evidence of at least 50 other stars similar to our Sun in the Universe. Because of its vast scale, it is impossible to imagine the size of the Universe, and according to scientists it is growing all the time. Somewhere there must be other planets, like our Earth, that can support life. But alien life forms could look so different that we would not recognize them.

We are still looking for those little green men...

Is There Life On Mars?

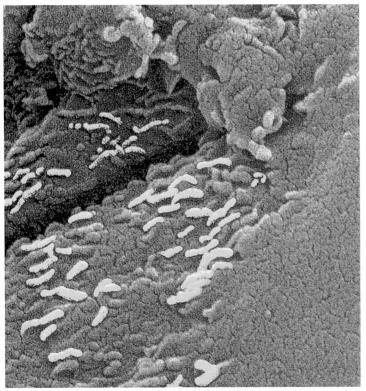

The question of whether there has ever been life on Mars was revived in 1996. Scientists using a powerful microscope to study a Martian meteorite found in Antarctica in 1984 noticed tiny "fossils" that looked like bacteria. These fossils are about 3.6 billion years old and may prove that Mars was once home to some form of life.

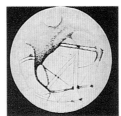

Martian Canals
In 1878, Giovanni Schiaparelli published the first of his maps of Mars. He called its long, straight lines "canals". Powerful telescopes have now proved that no such canals exist.

Map of Mars, 1890

Face on Mars
Photographs of the surface of Mars, taken in 1976, seem to reveal a giant face. It may be proof of an advanced civilization on that planet. Many people insist that the face is a signal to humans and shows that intelligent life exists on Mars.

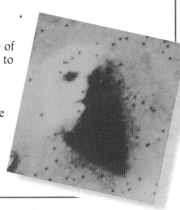

The Martian meteorite viewed under a microscope – the "fossils" have been coloured yellow

Photograph of Mars, 1976

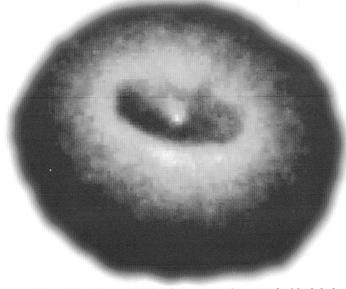

False colour image – the bright core may be part of a black hole

Into the Black Hole...
Astronomers believe that when huge stars reach the end of their lives their energy does not disappear. The force of their collapse is so great that they become as small as a pinpoint but weigh as much as a star. Although it is very small, this pinpoint has tremendous gravity. It sucks all surrounding matter into a kind of whirlpool, or "black hole", from which nothing can escape.

Hoax Moon Landing?

Buzz Aldrin on the *Apollo 11* mission

In spite of NASA film showing the lunar landing of 1969, some people refuse to believe that humans walked on the Moon. They claim that film of the *Apollo 11* mission was faked in a studio on Earth and that photographic clues have been left for the careful researcher. Some photographic experts find the pictures unconvincing and think that they look too posed.

TIME SLIPS

Time is a kind of one-way street, that does not allow us to travel backwards. Yet there are many examples of people who have apparently slipped into the past and of others who have had accurate glimpses of the future. Two of the most famous "time slips", at Versailles and Dieppe, took place in France.

Miss Eleanor Jourdain

Miss Charlotte Moberly

Queen Marie Antoinette

" The landscape became flat like a picture. "

Charlotte Moberly

STRANGE VISION AT VERSAILLES

On 10th August 1901, Miss Moberly, a headmistress of an Oxford college in the UK, and her friend Miss Jourdain experienced a time slip. While visiting the palace of Versailles, they were puzzled to see many people dressed in the old-fashioned clothing of the 18th century – including someone who looked exactly like Queen Marie Antoinette, the wife of King Louis XVI. Both women felt dreamy and depressed. Had they somehow slipped backwards in time to the 18th century?

The palace of Versailles, France

INDEX

Dieppe Harbour, France

Disaster at Dieppe

On 4th August 1951, two British women on holiday at Puys, in France, were kept awake by crashing noises that seemed to come from Dieppe harbour. They heard shouts, exploding shells, and the sound of aeroplanes. Both women believed that they had experienced a "time slip" back to a disastrous raid on Dieppe by British and Canadian forces on 19th August 1942.

Beast of Ben Macdui

When the mountaineer Professor Norman Collie went climbing alone on Ben Macdui in Scotland, UK, he heard crunching sounds behind him in the mist. He felt as though he was being followed by a giant and he fled down the mountain. For hundreds of years, people have heard the same footsteps and experienced feelings of panic.

Ben Macdui, Cairngorms, Scotland, UK

The chemical explosion at Flixborough, UK, 1974

News Flash from the Future

Twenty nine people died when a large chemical factory at Flixborough, Humberside, UK, exploded on 1st June 1974. Lesley Castleton learned about it – including the number of dead and injured – through a news flash on television at lunch time. However, according to the evening news, the explosion did not take place until 4.30 pm.